For Curious Minds

THE TEEN'S GUIDE TO PHYSICS

Aziz Chahir

For Curious Minds

Foreword

This booklet provides short definitions of the most well-known concepts in physics, accompanied by simple examples and small illustrations. It is designed for non-scientists and teenagers, making complex ideas easy to understand.

References

Basic Physics: A Self-Teaching Guide, 3rd Edition
by
Karl F. Kuhn
Frank Noschese

Images : Canva

Table of contents

Part One

Mathematics 17

Classical Physics 19

Newton's Laws of Motion 21

Law of Inertia 23
Law of Acceleration 24
Law of Action and Reaction 25
Understanding Velocity 26
The Law of Universal Gravitation 27
Apply the law of universal gravitation between the Earth and the Sun: 28

The Fascinating World of Optics 31

Optics 33
Light–Matter Interactions:
Refraction 36
Reflection 35
Diffraction 37
Interference 38

The Energy 39

Types of Energy:

- Kinetic Energy — 43
- Potential Energy — 45
- Thermal Energy — 48
- Radiant Energy — 51
- Electrical Energy — 54
- Nuclear Energy — 56

Thermodynamics 59

- System — 62
- Workon — 63
- Effeciency — 64

The laws of Thermodynamics :

- First law — 67
- Second law — 68
- third law — 69

Electromagnetism — 71

Maxwell's Equations: — 75

- Equation 1 — 79
- Equation 2 — 80
- Equation 3 — 81
- Equation 4 — 82

Part Two

Theoretical Physics — 87

- Special Relativity — 93
- General Relativity — 101
- Quantum mechanics — 113

Physics is often divided into two major areas:
- Classical Physics
- Theoretical Physics.

"Mathematics is the language in which God has written the universe"

Galileo Galilei

Mathematics is often called the language of physics because it's how we quantify and describe the physical world.

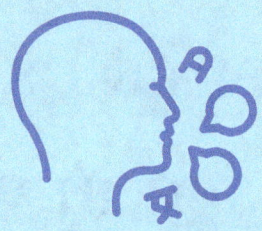

From measuring time and distance to calculating forces and energy, math is the tool that allows us to make sense of the universe.

Part One

Classical Physics

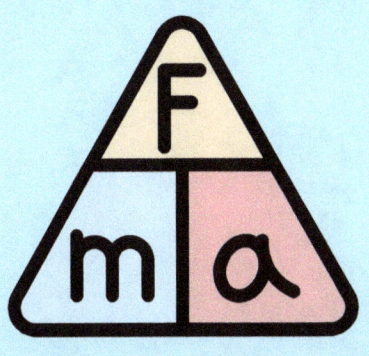

Newton's Laws of Motion

Law of Inertia:

An object at rest stays at rest,

An object in motion continues moving in a straight line at constant speed unless acted upon by an unbalanced force.

Picture a grocery cart rolling across the floor. It will keep moving until friction (an unbalanced force) slows it down.

Law of Acceleration:

Law of Acceleration: The acceleration of an object is directly proportional to the net force acting on it and inversely proportional to its mass:

$$acceleration = force/mass$$

The acceleration of an object depends on two things: how much force you use and how heavy the object is.

Simply put, the harder you push, the faster it moves unless it's really heavy, in which case it needs more force to achieve the same speed.

Law of Action and Reaction:

Law of Action and Reaction: For every action, there's an equal and opposite reaction.

$Fa = -Fb$

Push on a wall, and the wall pushes back with equal force, even though it doesn't move.

Understanding Velocity:

Velocity isn't just speed—it's speed with direction. While speed tells you how fast something is moving, velocity also tells you which way it's going.

It's like the difference between knowing a car is going 60 km/h and knowing it's going 60 km/h to the north.

The Law of Universal Gravitation:

Isaac Newton's law of universal gravitation explains that every object in the universe attracts every other object

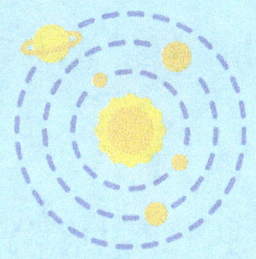

with a force proportional to their masses and inversely proportional to the square of the distance between them.

$$F = G \cdot \frac{m_1 \cdot m_2}{r^2}$$

The bigger and closer two objects are, the stronger the gravitational pull between them.

Apply the law of universal gravitation between the Earth and the Sun:

To apply the law of universal gravitation between the Earth and the Sun, we'll use Newton's formula:

$$F = G \cdot \frac{m_1 \cdot m_2}{r^2}$$

F = gravitational force between the two bodies

G = gravitational constant

m1 = mass of the first body (Sun)

m2 = mass of the second body (Earth)

r = distance between the centers of the two bodies

N = One Newton is the force needed to accelerate a mass of one kilogram at a rate of one meter per second squared.

G 6.7×10^{-11}

- It's a fundamental physical constant used in calculations involving gravity.
- The units are $N(m/kg)^2$ (Newton square meters per kilogram squared).

Mass of Sun 2×10^{30} kg

Mass of Sun 2×10^{30} kg

r (the average distance between the Earth and the Sun) 1.5×10^{11} m

$F = G * (M1 * M2) / r^2$. Do your calculations

Therefore, the gravitational force between the Earth and the Sun is approximately 3.5415×10^{22} N.

If you don't understand these calculations, don't worry. They're just to give you a general idea.

The exact numbers aren't crucial for a basic understanding. The key is grasping that there's a strong gravitational pull between the Earth and Sun, and this is what maintains Earth's orbit.

The Fascinating World of Optics

Optics

Optics is the branch of physics that explores how light behaves when it interacts with different materials.

Whether light is reflecting off a mirror, bending through a lens, or creating a rainbow, optics helps us understand and manipulate the properties of light.

Light–Matter Interactions:

Reflection:

Reflection: Light bounces off surfaces

which is why we see objects when light reflects off them and into our eyes.

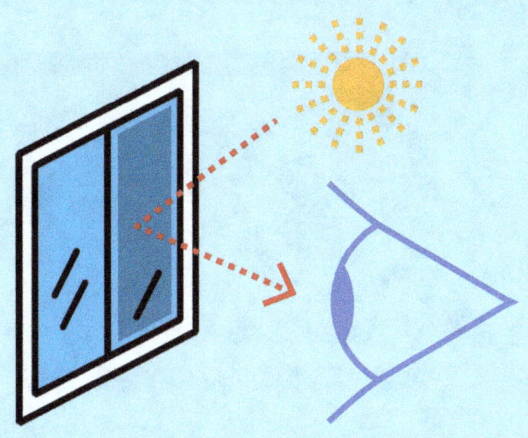

Refraction:

Refraction: Light bends when it passes through different materials, like water or glass,

which is how lenses can focus light and magnify images.

Diffraction:

Light spreads out when it encounters obstacles,

creating patterns like the colorful edges of shadows.

Interference:

Interference: When light waves meet, they can combine to create bright and dark patterns,

explaining phenomena like holograms and the iridescent colors in soap bubbles.

The Energy That Powers the Universe

Energy is a fundamental concept in physics, and it comes in many forms:

Types of Energy:

Kinetic Energy:

Kinetic Energy: The energy of motion—everything that moves has kinetic energy.

The amount of kinetic energy depends on two main factors:

- Mass: Heavier objects have more kinetic energy.

- Speed: Faster objects have more kinetic energy.

Some everyday examples of kinetic energy include:
- A car driving down the road

- A ball rolling across the floor

- A person walking or running

- Water flowing in a river

This energy can be transferred to other objects through collisions or converted into other forms of energy.

Potential Energy:

Potential energy is stored energy that has the potential to do work.

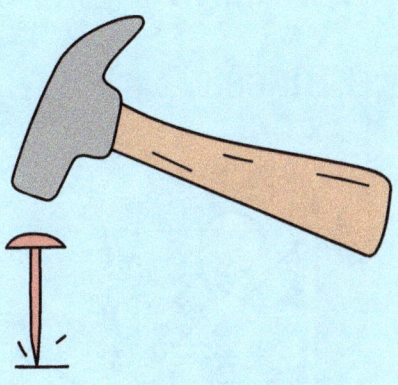

It's energy an object possesses due to its position or state, rather than its motion.

Key points about potential energy include:

Position matters: Objects have gravitational potential energy based on their height and mass.

Configuration: Some objects store energy based on how they're arranged or shaped.

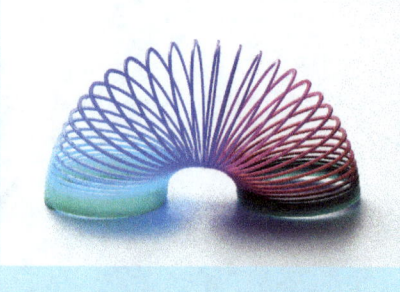

Common examples of potential energy:

A book on a high shelf (gravitational potential energy)

A stretched rubber band (elastic potential energy)

A compressed spring (elastic potential energy)

A drawn bow (elastic potential energy)

Food (chemical potential energy)

Thermal Energy:

Thermal energy is the energy associated with the temperature of an object or substance.

It's related to the motion and vibration of atoms and molecules.

Key points about thermal energy include:

Temperature: Higher temperature means more thermal energy.

Particle motion: Faster-moving particles indicate more thermal energy.

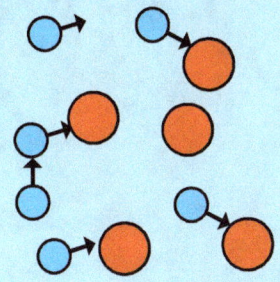

Examples of thermal energy in everyday life::

The warmth you feel from a hot cup of coffee.

The warmth of the sun on your skin

Radiant Energy:

Radiant energy is energy that travels in waves, specifically electromagnetic waves.

It includes visible light and other forms of radiation.

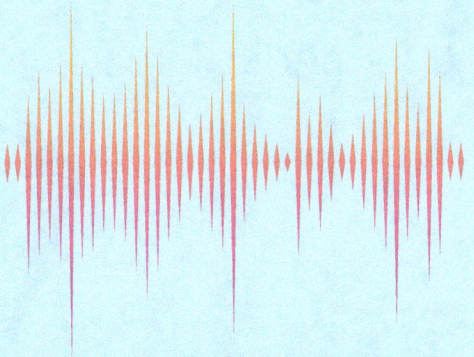

Key points about radiant energy:

Light from a lamp allowing you to read

Radio waves carrying music to your car radio

Microwaves heating your food

X-rays used in medical imaging

Radiant energy can be absorbed,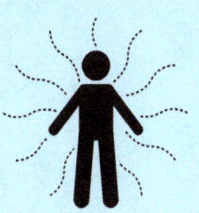

reflected,  or transmitted by

different materials.

It's a crucial form of energy transfer,

especially from the sun to Earth.

Electrical Energy:

Electrical energy is the energy carried by moving electric charges, typically electrons.

It's a versatile form of energy that we use extensively in modern life.

Key points about electrical energy:

Flow of charges: It's created by the movement of electrons through a conductor.

Easily converted: Can be readily transformed into other forms of energy.

Nuclear Energy:

Nuclear energy is the energy stored within the nucleus (core) of atoms.

It's released through nuclear reactions, either fission (splitting atoms)

or fusion (joining atoms).

Key points about nuclear energy:

Extremely powerful: Releases far more energy than chemical reactions.

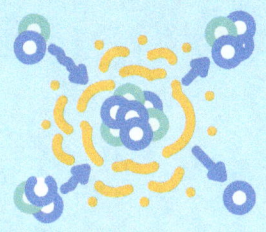

Occurs naturally: In stars like our sun, and in radioactive elements on Earth.

Thermodynamics:

Thermodynamics is the branch of physics that deals with heat,

temperature,

and their relation to energy and work. It describes how thermal energy is converted to and from other forms of energy and how it affects matter.

Key concepts in thermodynamics:

Energy transfer: How heat moves between objects or sysytem

A "system" in thermodynamics can be almost anything we choose to focus on. Here are some simple examples:
1. A pot of boiling water
2. An ice cube melting in a glass
3. The engine of a car

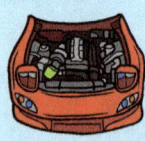

Work:

<u>How energy can be used to cause physical changes.</u>

Here are some simple examples:

Lifting an object: Energy is used to move it against gravity.

Compressing a gas: Like pumping air into a tire.

Stretching a spring: Energy is stored in the spring's new shape.

Efficiency:

The relationship between energy input and useful output.

Here are some simple examples:

Car Engine: When you fill up a car with gasoline, not all of that fuel's energy moves the car forward. Some energy is lost as heat and noise. If your car engine is 25% efficient, that means 25% of the energy in the gasoline actually powers the car, while 75% is lost.

Light Bulb: An old incandescent light bulb uses electricity to produce light, but it also gets hot. If it's only 10% efficient, 10% of the electrical energy becomes light, while 90% is wasted as heat.

The laws of thermodynamics

first law:

Energy cannot be created or destroyed, only converted from one form to another. Here is simple example:
When you pedal a bicycle, your muscles convert the chemical energy stored in your body (from the food you eat) into mechanical energy. This mechanical energy turns the wheels, making the bike move forward. As you ride, some of the energy is also converted into heat (you get warmer) and sound (the noise of the tires on the road).

Second law:

Heat naturally flows from hot to cold, not the reverse.

Here is simple example:

When you put an ice cube in a glass of warm water, the heat from the water naturally flows into the colder ice cube.

This causes the ice to melt and the water to coo down. The heat always moves from the warmer water to the colder ice, not the other way around.

Third law:

It's impossible to reach absolute zero temperature.
Here is simple example:

Imagine you're trying to cool something down to absolute zero (which is –273.15°C or –459.67°F). No matter how much you try to remove the heat, you can never quite get rid of all of it. There will always be a tiny bit of heat left, and the closer you get to absolute zero, the harder it becomes to remove that last bit. So, absolute zero is like a finish line you can never actually cross. IMPOSSIBLE

Electromagnetism

The Force Behind Our Modern World

Electromagnetism is one of the four fundamental forces of nature.(*) Electromagnetism is the force responsible for the interactions between electrically charged particles, This electromagnetic force is what enables atoms to bond together, forming molecules and solid materials in our everyday world.

It's the "invisible glue" that keeps most solid materials together.

....

(*)
- Electromagnetic force
- Strong nuclear force
- Weak nuclear force
- Gravitational force

It's the force that involves the interaction between electrically charged particles and magnetic fields.

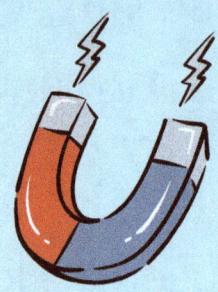

This force is responsible for many of the phenomena we observe in our daily lives.

Maxwell's Equations

Maxwell's Equations are a set of four fundamental equations in physics that describe how electricity and magnetism work. They were formulated by physicist James Clerk Maxwell in the 19th century.

In very simple terms, these equations tell us:

Electric charges produce electric fields.

Imagine you have a balloon that you've rubbed against your hair, giving it a negative electric charge. Now, if you bring this balloon close to small pieces of paper, you'll see the paper get attracted to the balloon and stick to it:

1. The charged balloon creates an electric field around itself.
2. This electric field exerts a force on the nearby paper.
3. The force causes the paper to move towards the balloon.

Magnetic poles don't exist in isolation.

1. Magnetic poles: When we talk about magnets, we usually think of them as having two ends – a north pole and a south pole. These ends are called magnetic poles.
2. You can't have just a north pole or just a south pole by itself. Every magnet always has both a north and a south pole.

If a bar magnet is broken into two or more pieces, each of them will have a north pole and a south pole.

Changing magnetic fields create electric fields.

- Magnetic fields: Think of the area around a magnet where you can feel its pull. That's a magnetic field.

- When a magnetic field moves or gets stronger or weaker, it can cause an electric field to appear.

Electric currents and changing electric fields create magnetic fields.

Electric Currents and Magnetic Fields:
- When an electric current flows through a wire, it creates a magnetic field around the wire. You can imagine it like an invisible ring of force circling around the wire. The stronger the current, the stronger the magnetic field.

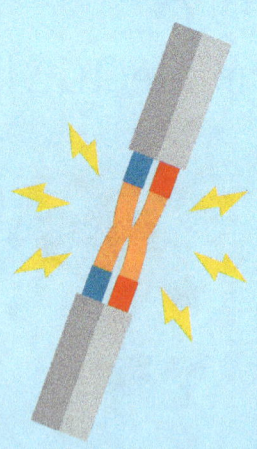

Changing Electric Fields and Magnetic Fields:

- When an electric field changes over time (like when you turn a switch on or off),

it can create a magnetic field.

why electromagnetism is so important in our modern world:

- Electricity: The flow of electric charge powers almost all of our technology, from smartphones to refrigerators

- Electronics: All electronic devices, including computers, televisions, and radios, rely on the principles of electromagnetism..

- Communications: Electromagnetic waves are used for wireless communication, including radio, TV, Wi-Fi, and mobile phones.

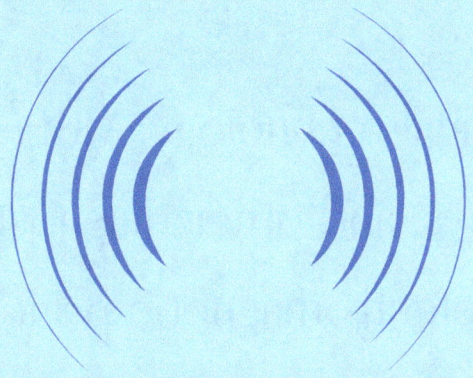

- Lighting: Electric lights, from traditional bulbs to LEDs, work because of electromagnetic principles.

- Transportation: Electric motors in cars, trains, and other vehicles use electromagnetic forces.

- Medical technology: MRI machines, X-rays, and other diagnostic tools use electromagnetic principles.

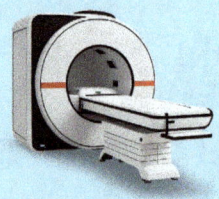

- Renewable energy: Solar panels and wind turbines generate electricity using electromagnetic induction.

Part Two

Theoretical Physics

88

Theoretical physics is the branch of physics that uses mathematical models and abstractions to explain, predict, and understand natural phenomena.

It focuses on developing theories and frameworks that describe the fundamental laws governing the universe.

Relativity & Quantum mechanics

are two of the most important branches of theoretical physics because they form the foundation of our understanding of the physical world:

Relativity

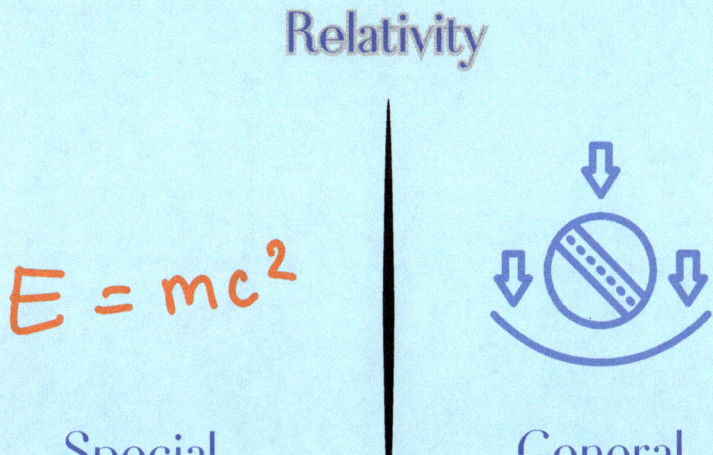

Special | General

Redefines our understanding of space, time, and gravity, explaining the behavior of objects at high speeds and in strong gravitational fields.

Quantum mechanics

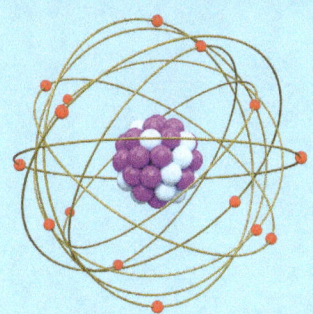

Describes the behavior of particles at the atomic and subatomic levels.

Special Relativity

Special relativity!

Is a theory developed by Albert Einstein that describes how objects behave when they move at speeds close to the speed of light.

The speed of light is approximately 299,792 kilometers per second in a vacuum.

In simple terms, special relativity shows that our common sense ideas about time, space, and motion don't apply when things are moving very fast, especially close to the speed of light.

It changed our understanding of the universe and has been confirmed by many experiments.

Key Ideas of Special Relativity:

Speed of Light:

The speed of light in a vacuum is always the same, no matter how fast you're moving or where you are in the universe.

Upper Speed Limit: According to special relativity, nothing can travel faster than the speed of light.

<u>Time and Space:</u> Time and space are not absolute. This means that time can pass differently for two people depending on how fast they are moving relative to each other.

A clock moving very fast will tick more slowly compared to a clock that is at rest. This effect is called <u>time dilation.</u>

$\underline{E=mc^2}$: This famous equation shows that energy (E) and mass (m) are related; a small amount of mass can be converted into a large amount of energy.

This is why nuclear reactions can release so much energy.

100

General Relativity

general relativity

General relativity is a theory of gravity developed by Albert Einstein. It explains that massive objects, like planets and stars, curve space and time around them, causing other objects to move along these curves. This is why planets orbit the sun and why light bends near massive objects.

Key Ideas of General Relativity:

Gravity as Curvature of Spacetime:

In general relativity, gravity is not described as a force, as it was in Newton's theory.

Instead, Einstein proposed that massive objects like the Earth and the Sun cause a curvature in the fabric of spacetime (a four-dimensional combination of space and time).

Imagine spacetime as a stretched rubber sheet. When you place a heavy object on it, like a bowling ball, the sheet bends.

Smaller objects, like marbles, will roll towards the bowling ball because of this curvature.

Similarly, planets orbit the Sun because the Sun's massive presence bends spacetime around it, and the planets follow this curvature.

Spacetime:

pacetime is a concept that combines the three dimensions of space (length, width, height) with the fourth dimension of time.

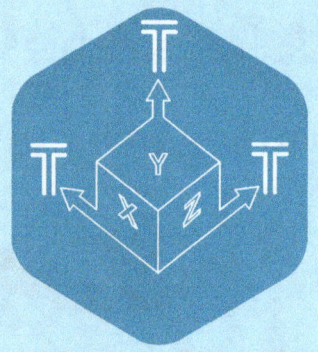

General relativity shows that massive objects distort this spacetime, and this distortion is what we perceive as gravity.

Key Predictions:

Gravitational Time Dilation:

Time runs slower in stronger gravitational fields. For example, time passes more slowly closer to the Earth's surface than it does further away.

A clock moving very fast will tick more slowly compared to a clock that is at rest..

Gravitational Waves:
Ripples in spacetime caused by accelerating massive objects, like merging black holes.

These waves were predicted by Einstein and first directly detected in 2015.

Black Holes:

Regions of space where gravity is so strong that not even light can escape. Black holes are a direct prediction of general relativity.

Most black holes are born from the death of massive stars. When these giant stars run out of fuel, they explode in a spectacular event called a supernova. What's left behind can collapse into a super-dense black hole.

Impact of General Relativity:

General relativity has been confirmed by many experiments and observations,

such as the bending of light around massive objects (gravitational lensing)

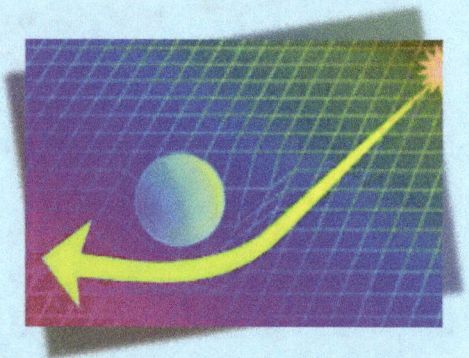

such as the bending of light around massive objects (gravitational lensing)

Geometric property of spacetime

In essence, general relativity gives us a deeper and more accurate understanding of gravity, showing it as a geometric property of spacetime itself rather than a force acting at a distance.

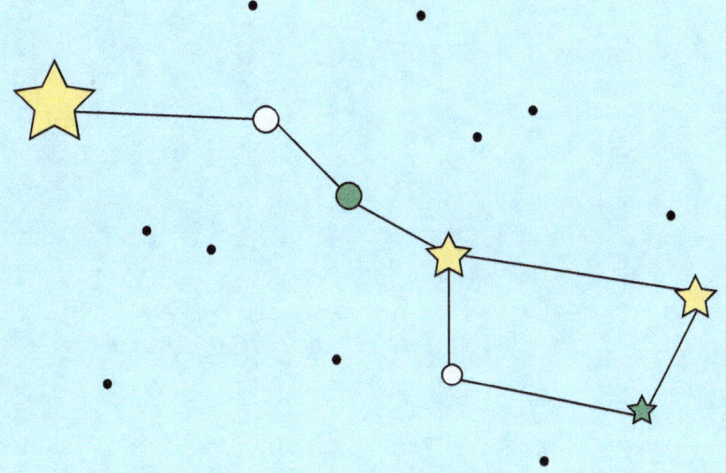

112

Quantum mechanics

Quantum mechanics

Quantum mechanics is the branch of physics that explains how very tiny things, like atoms and subatomic particles (electron, photon,...) behave.
Unlike everyday objects, these tiny particles don't follow the same rules as larger objects; they act in ways that can seem strange and counterintuitive.

Key Ideas of Quantum Mechanics:

Wave-Particle Duality:

- Tiny particles, like electrons and photons (particles of light), can behave both like particles and like waves.

For example, light can spread out like a wave, but it also comes in packets called photons, like tiny particles.

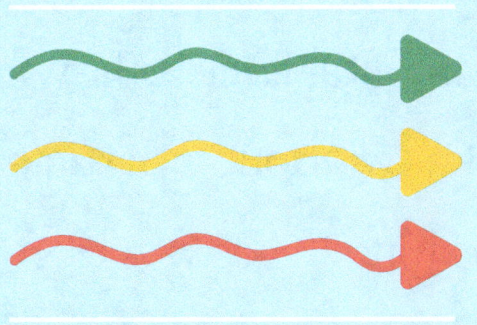

Uncertainty Principle:

- The Heisenberg Uncertainty Principle says that we can't know both the exact position and the exact speed of a particle at the same time.

- The more precisely we know one, the less precisely we can know the other. This means there's a fundamental limit to how much we can know about a particle's behavior.

Superposition:

- Particles can exist in multiple states at once until they are observed. For example, an electron can be in more than one place at the same time, but when we measure its position, it "chooses" a single place.

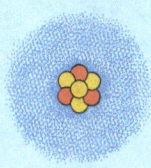 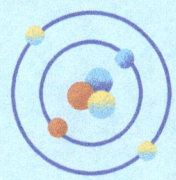

- This idea is famously illustrated by Schrödinger's cat, a thought experiment where a cat in a box is both alive and dead until someone opens the box and checks.

Entanglement:

- Quantum entanglement is a phenomenon where particles become linked, so that the state of one particle instantly influences the state of another, no matter how far apart they are.

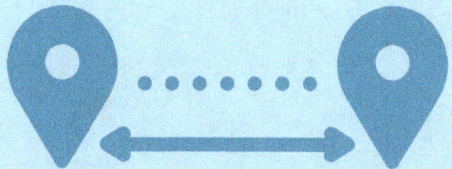

- This "spooky action at a distance," as Einstein called it, has been confirmed by experiments and is a key feature of quantum mechanics.

Why Quantum Mechanics Matters:

- **Explains the Atomic World:** Quantum mechanics helps us understand how atoms and molecules work, which is crucial for chemistry, biology, and material science.

- **Technological Applications:** It's the foundation for many modern technologies, including semiconductors (which power computers), lasers, and MRI machines..

<u>Fundamental Nature:</u>

Quantum mechanics challenges our everyday understanding of reality, showing that at the smallest scales, nature is fundamentally probabilistic, not deterministic.

Conclusion:

This guide has offered a glimpse into the fascinating world of physics, introducing you to some of its core concepts and ideas. From the grand scales of the cosmos to the minute realm of quantum mechanics, we've explored how physics helps us understand and describe the universe around us.

Remember, what you've read here is just the tip of the iceberg. Physics is a vast and ever-evolving field, with new discoveries constantly reshaping our understanding of reality. The concepts we've covered – like energy, matter, space-time, and fundamental forces – are foundational, but there's always more to learn and explore.

www.ingramcontent.com/pod-product-compliance
Lightning Source LLC
Chambersburg PA
CBHW050309230526
45471CB00005B/2093